AQA Mathematics

Foundation (Linear)

Exam Skills Question Book

GCSE

Series Editor

Paul Metcalf

Series Advisor

Andy Darbourne

Lead Authors

Sandra Burns
Shaun Procter-Green
Margaret Thornton

Authors

Tony Fisher
June Haighton
Anne Haworth
Gill Hewlett
Steve Lomax
Jan Lucas
Andrew Manning
Ginette McManus
Howard Prior
David Pritchard
Dave Ridgway
Kathryn Scott
Paul Winters

Nelson Thornes

This edition published in 2013 by:
Nelson Thornes Ltd
Delta Place
27 Bath Road
CHELTENHAM
GL53 7TH
United Kingdom

13 14 15 16 / 10 9 8 7 6 5 4 3 2

A catalogue record for this book is available from the British Library

ISBN 978 1 4085 2148 9

Cover photographs: Purestock/Getty and Steve Debenport/Getty
Page make-up by Tech-Set Limited, Gateshead
Printed in China by 1010 Printing International Ltd

Photograph acknowledgements

Fotolia: pages 10, 18, 26, 27, 29, 44
Getty Images: Joe Patronite page 9

Contents

Welcome to GCSE Mathematics

This book has been written by teachers who not only want you to get the best grade you can in your GCSE exam, but also to enjoy maths. Together with Books 1 and 2, it offers you the opportunity to put your skills and knowledge into practice.

In the exam you will be tested on the Assessment Objectives (AOs) below. Ask your teacher if you need help to understand what these mean. The questions in this book focus on AO2 and AO3 type questions.

AO1   recall and use your knowledge of the prescribed content.

AO2   select and apply mathematical methods in a range of contexts.

AO3   interpret and analyse problems and generate strategies to solve them.

Each chapter starts with a list of topics that will be needed for the exercise that follows. The questions will draw on one or more of the topics listed, so you will need to decide what methods to use to solve the problems.

How to use this book

To help you unlock blended learning, we have referenced the activities in this book that have additional online coverage in *Kerboodle* by using this icon:

The icons in this book show you the online resources available from the start of the new specification and will always be relevant.

In addition, to keep the blend up-to-date and engaging, we review customer feedback and may add new content onto *Kerboodle* after publication!

You will see the following features throughout this book.

 The bars that run alongside questions in the exercises show you what grade the question is aimed at. This will give you an idea of what grade you're working at. Don't forget, even if you are aiming at a Grade C, you will still need to do well on the Grades G–D questions.

Hint

These are tips for you to remember whilst learning the maths or answering questions.

Study tip

Hints to help you with your study and exam preparation.

Bump up your grade

These are tips, giving you help on how to boost your grade, especially aimed at getting a Grade C.

Section 1

Fractions

Decimals

Collecting data

Percentages

Ratio and proportion

Statistical measures

Representing data

Scatter graphs

Probability

All these topics will be tested in this chapter and you will find a mixture of problem solving and functional questions. You won't always be told which bit of maths to use or what type a question is, so you will have to decide on the best method, just like in your exam.

Example: The length of 50 films is given in the table.

Length of film (min)	Frequency
$70 \leqslant l < 80$	19
$80 \leqslant l < 90$	22
$90 \leqslant l < 100$	7
$100 \leqslant l < 110$	2

Copy the graph below and draw a suitable diagram to show the data.

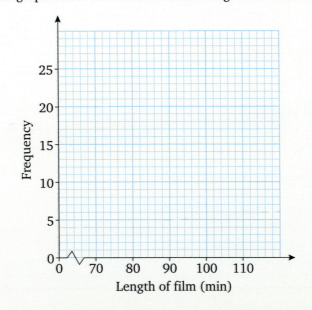

(4 marks)

Solution:

Study tip

- Before you start, think about the type of data in the question.
- Time is continuous data so you must choose a diagram for continuous data.
- Choosing an inappropriate diagram will probably score zero.
- In an exam, the graph paper will usually have axes drawn and labelled.

For this continuous data, the choices are to draw a frequency diagram or a frequency polygon.

Frequency diagram

For a frequency diagram, the height of each bar represents the frequency for each class interval. For continuous data, the bars are joined together.

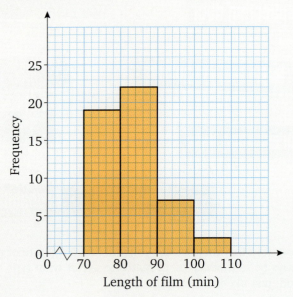

Frequency polygon

For a frequency polygon, plot the frequency values at the midpoint of each class interval. The points should be joined to create the frequency polygon.

For example, the midpoint of the first class is $\dfrac{70 + 80}{2} = 75$

Study tip

There is no need to draw any lines before the first point or after the final point.

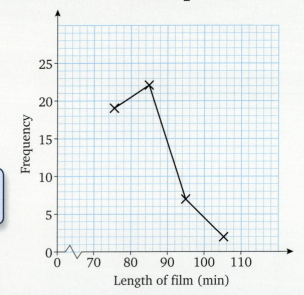

Mark scheme

- 1 mark for drawing an appropriate diagram.
- 2 marks for drawing the bars to the correct height, or plotting the points – you lose one mark for making one mistake.
- 1 mark for drawing bars without gaps or for joining the points in a frequency polygon.

Example: This spinner has six equal sections.

a Write down the word that describes how likely it is that the spinner does each of the following.

    i Lands on yellow

       certain   even chance   impossible

    ii Lands on blue

       certain   even chance   impossible

b Here are two spinners A and B.

        Spinner A   Spinner B

Which spinner is more likely to land on green?

Explain your answer.

(4 marks)

Solution: **a** **i** <u>Impossible</u> as there are no yellow sections to land on.

    ii <u>Even chance</u>.

       There are three blue sections.

       Probability of blue $= \frac{3}{6} = \frac{1}{2}$

       The probability $\frac{1}{2}$ is the same as even chance.

b For spinner A, probability of green $= \frac{3}{4}$

    For spinner B, probability of green $= \frac{4}{6} = \frac{2}{3}$

    To compare the two probabilities it is helpful to convert to percentages.

    For spinner A, probability of green $= \frac{3}{4} = 75\%$

    For spinner B, probability of green $= \frac{4}{6} = \frac{2}{3} = 66\frac{2}{3}\%$

    So spinner A is more likely to land on green.

> **Study tip**
>
> You could also convert the fractions to equivalent fractions with the same denominator:
>
> $\frac{3}{4} = \frac{9}{12}$ and $\frac{2}{3} = \frac{8}{12}$

> **Mark scheme**
> - 1 mark for part **a i** correct.
> - 1 mark for part **a ii** correct.
> - 1 mark for finding probabilities in part **b**.
> - 1 mark for saying spinner A is more likely **and** offering a correct explanation.

Questions

1 **a** Write 0.75 as a fraction.

    b Write $\frac{9}{100}$ as a percentage.

    c Write two fifths as a decimal.

    d Write $\frac{11}{100}$ as a decimal.

    e Write down a fraction between $\frac{1}{5}$ and $\frac{2}{5}$

G
F
E

G F E **2** Here are four statements.

Decide whether each one is **always true**, **sometimes true** or **never true**.

In each case explain your choice of answer.

a In a bar chart, the bars are different widths.

b In a bar chart, the bars are drawn vertically.

c In the key for a pictogram, each symbol represents 10.

d The symbols used in a pictogram are all the same size.

e In a pie chart, the size of the angle represents the frequency.

f In a pie chart, the area of the sector represents the frequency.

G E **3** The table shows the sales of different types of tea.

Tea	Ordinary	Apple	Blackcurrant	Lemon	Other
Frequency	18	8	4	5	1

Show this information as:

a a pictogram   **b** a bar chart   **c** a pie chart.

F E **4** A computer costing £900 has its price reduced by £150 in a sale.

a **i** By what fraction is the price reduced?

ii What fraction of the sale price is the reduction?

b The sale price of another computer in the sale was £600 after a reduction of $\frac{1}{4}$
What was the original price?

E **5** **a** Ellie says that $0.2 \times 0.4 = 0.8$
Explain why Ellie is wrong.

b Copy this calculation and fill in the missing number.

$0.3 \times \boxed{} = 6$

6 Amir is given a pay rise of £500.

He says this is only a 2% rise on his current pay.

How much will Amir earn after the pay rise?

G D C **7** The chart shows the number of houses on an estate in 2004 and 2006.

a How many houses were built between 2004 and 2006?

b The number of houses built between 2006 and 2008 was only $\frac{2}{3}$ of the number built between 2004 and 2006.
Copy the chart and draw in the bar for 2008.

c In 2003, the builder was given the following target:

> ratio of the number of houses in 2004 : number of houses in 2010 must not be less than 1 : 2.1

Between 2008 and 2009 he built 7 houses.

Do you think the builder is likely to meet his target?

You **must** show working to justify your answer.

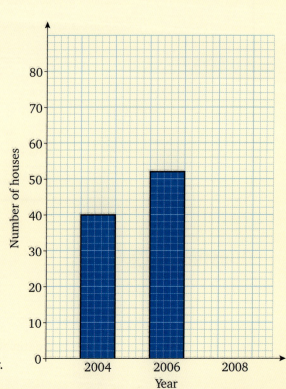

8 A gym club has 18 male and 12 female members.

    a Julie says that the fraction of female members is $\frac{12}{18}$
Is Julie correct? Give a reason for your answer.

    b 20 new members join the club.
The ratio of male to female members is now $2:3$
How many of the new members were male?

9 Zafeer runs 400 m races.

He has taken part in 9 races so far and his mean time is 50.08 seconds.

His aim is to achieve a mean time under 50 seconds.

He has one race left to run.

What time must he run if he is to achieve his aim?

10 Prita starts with a number between 4 and 6.

She adds 0.2 to her number and then doubles the answer.

She repeats these two steps a number of times until she gets to 50.

What number does Prita start with?

11 Copy the diagram. Then connect each of the following to its proper description.
The first one has been done for you.

The age of lions at London Zoo
People's favourite cake at a show
Points scored in 10 darts matches
The weight of 12 newborn puppies
The favourite building of people in Britain
The average speed of train journeys into London
The number of hours spent driving per day
A person's shirt collar size
The cost of bread
Rainfall at a seaside resort

Quantitative and discrete
Qualitative
Quantitative and continuous

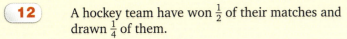

12 A hockey team have won $\frac{1}{2}$ of their matches and drawn $\frac{1}{4}$ of them.

They have lost six matches.

They are awarded two points for a win,
one point for a draw
and no points for a loss.

How many points have the team been awarded?

13 A recipe lists the following quantities of ingredients to make six scones.

    a List the ingredients needed to make 12 scones.

    b How much flour is needed for 15 scones?

    c Write the ratio of flour : sultanas : butter : sugar in its simplest form.

Six scones ...

Self-raising flour	240 g
Salt	0.5 teaspoon
Sultanas	75 g
Butter	40 g
Caster sugar	25 g
Egg	1 large
Milk	20 ml

D

14   Andy and Roger have played each other at tennis 40 times.

Roger has won 60% of the games.

Andy has won three times more games on a hard court than on a grass court. 35% of the games have been played on a grass court.

All other games have been played on a hard court.

a   Copy and complete the table.

	Number of wins for Andy	Number of wins for Roger
Played on grass court		
Played on hard court		

b   Andy and Roger play their next match tomorrow.

If it is raining it will be played on a hard court.

If it is fine it will be played on a grass court.

What would Andy prefer?

Give a reason for your answer.

15   In football, the referee shows a player a yellow card for a bad foul and a red card when the player is sent off.

The scatter graph shows the number of red and yellow cards awarded by 12 referees labelled A to L in the Premier League for the 2008–2009 season.

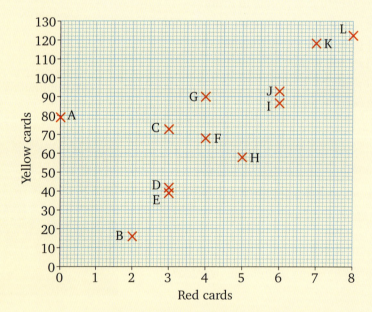

a   One referee showed 8 times more yellow cards than red cards.

Which referee was this?

b   The mean number of cards per match shown by G was 4.087 to 3 decimal places.

How many matches did G referee?

c   Angie says, 'The more yellow cards a referee shows, the more red cards the referee shows.'

Comment on this statement.

16   A spinner has 10 equal sections.

Each section is red, blue or green.

All three colours are used at least once.

The probability of the spinner landing on green is the same as the probability of the spinner landing on blue.

The colour that the spinner is most likely to land on is red.

Show **two** ways that the spinner could be labelled.

17   The table shows the engine sizes and maximum speeds of eight cars.

Engine size (cc)	Maximum speed (mph)
1100	80
1800	125
2900	142
1400	107
1300	96
1000	85
2500	135
2000	131

a   Draw a scatter graph to show these results.

b   Describe the relationship between a car's engine size and its maximum speed.

c   Use a line of best fit to estimate:

    i   the maximum speed of a car with an engine size of 1500 cc

    ii   the engine size of a car whose maximum speed is 150 mph.

d   Explain why your last answer might not be accurate.

18   Students at a college are asked to choose their favourite colour.

Their choices are shown in the pie chart.

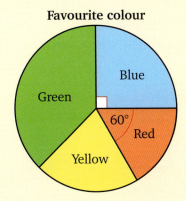

Favourite colour

A total of 45 students choose the colour blue.

Twice as many students choose green as yellow.
How many students choose green?

D
C

19 Mel pins up four posters on a noticeboard.

Each poster is 20 cm wide.

The posters overlap by 0.8 cm.

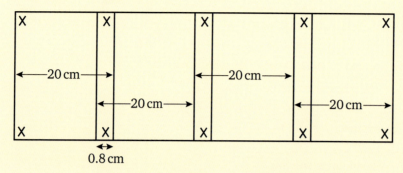

0.8 cm

Work out the total width of the four posters on the noticeboard.

20 In a bingo club, 87% of the members are female.

a What percentage of members are male?

b The manager sends a voucher to every fourth name on the membership list.

He sends out 155 vouchers altogether.

What is the least number of members that the club has?

c The manager is thinking about closing the club on Mondays.

He gives a questionnaire about this to the first 10 males and the first 10 females to enter the club one Friday.

Is this a good sample to choose?

Give **two** reasons for your decision.

C

21 The mean of five numbers is 27.6

When the numbers are arranged in order the first two numbers are the same. The difference between each of the other numbers is always 2.

Work out the median.

22 An ordinary pack of cards has four suits called clubs, diamonds, spades and hearts. Each suit has an Ace, number cards 2–10 and a Jack (J), Queen (Q) and King (K). There are 52 cards in total.

The table shows the frequency distribution after choosing a card at random 520 times from a complete pack.

Results from 520 random choices		
Picture card (J, Q, K)	Ace	Number card
119	44	357

(left label: Frequency distribution)

a What is the relative frequency of getting an Ace?

b Which of the results in the table is the closest to the result predicted by theoretical probability?

23 The table shows the time spent waiting at a clinic.

Time, t (minutes)	Frequency
$0 \leqslant t < 10$	4
$10 \leqslant t < 20$	14
$20 \leqslant t < 30$	4
$30 \leqslant t < 40$	1

Draw a suitable diagram to display the data.

24 A new medicine to cure the common cold is tested.

Eighty people are divided into two equal groups, A and B.

Only one of the groups is given the new medicine.

The time taken by the people in each group to overcome their cold is recorded.

Number of days to overcome cold	Group A frequencies	Group B frequencies
1	0	1
2	5	7
3	15	15
4	13	
5	7	

The doctor forgot to fill in the last **two** rows for Group B.

He says that the mean of both groups is the same.

Show clearly why the missing frequencies must be 3 and 14.

Examination-style questions Ⓚ

1 A student recorded the time, in minutes, that 50 people spent in the library.

Time, t (minutes)	Frequency
$0 < t \leqslant 10$	2
$10 < t \leqslant 20$	8
$20 < t \leqslant 30$	20
$30 < t \leqslant 40$	12
$40 < t \leqslant 50$	8

Calculate an estimate of the mean number of minutes spent in the library.

(4 marks)

AQA 2008

2 David, Gareth and Kerry share out the contents of a jar of 600 sweets.

David receives $\frac{1}{4}$ of the sweets.

Gareth receives $\frac{5}{8}$ of the sweets.

What fraction of the sweets is left in the jar for Kerry?

(4 marks)

AQA 2007

3   Linda works at Shopsave.
She wonders if she would earn more working at Superspend.
Here is some information about Shopsave and Superspend.

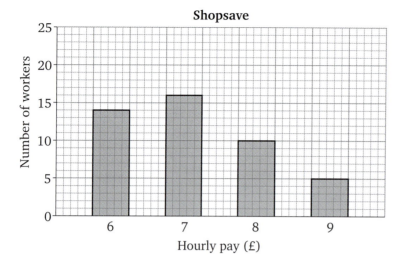

Superspend	
Hourly pay (£)	Number of workers
6	18
7	10
8	6
9	4
10	2

a   For the workers at Shopsave:

   i   How many workers are paid £6 per hour?   *(1 mark)*

   ii   What fraction of workers are paid £9 per hour?   *(2 marks)*

   iii   What is the range of their hourly pay?   *(1 mark)*

b   The mean hourly pay at Shopsave is £7.13.

   Show that the mean hourly pay at Superspend is £7.05.   *(3 marks)*

c   Draw a dual bar chart to show the data for Shopsave and Superspend on the same graph.   *(3 marks)*

d   Compare the hourly pay at the two shops.

   i   Give **one** reason why Linda might apply for a job at Superspend.   *(1 mark)*

   ii   Give **one** reason why Linda might stay at Shopsave.   *(1 mark)*

AQA 2008

4   Bags of carrots are labelled as weighing 500 grams.
The diagram shows the actual weight, to the nearest gram, of a sample of 25 bags.

```
49 │ 1  3  6  7  7  8  8  9  9
50 │ 0  0  0  1  1  2  2  3  5  7
51 │ 0  4  4  8  9
52 │ 3
```

Key: …… **│** …… represents …… grams

a   Copy and complete the key.   *(1 mark)*

b   Find the median weight for these bags of carrots.   *(1 mark)*

c   The company who pack the carrots say that:

> More than half of the bags are overweight.
> All bags are within 2% of 500 grams.

Investigate these statements using this sample of bags.   *(4 marks)*

AQA 2008

Before attempting this chapter, you will need to have covered the following topics:

Types of numbers

Fractions

Working with symbols

Equations and inequalities

Indices

Formulae

Sequences

Decimals

Coordinates

Percentages

Graphs of linear functions

Ratio and proportion

Real-life graphs

All these topics will be tested in this chapter and you will find a mixture of problem solving and functional questions. You won't always be told which bit of maths to use or what type a question is, so you will have to decide on the best method, just like in your exam.

Example: A dressmaker has orders to make three skirts.

One of the skirts is size 8, one size 10 and one size 14.

The dressmaker needs to buy the fabric to make the skirts.

The table shows the length of fabric needed.

Size	8	10	12	14	16	18
Length of fabric	1.9 m	1.9 m	1.9 m	2.2 m	2.2 m	2.2 m

The fabric costs £5.50 per metre.

How much does the dressmaker have to pay for the fabric to make these skirts?

(3 marks)

Solution: Total length of fabric: $1.9 + 1.9 + 2.2 = 6$

So 6 metres of fabric is needed.

Cost of fabric: $6 \times 5.50 = 33$

So the fabric costs £33.00.

> **Study tip**
>
> In order to obtain the method marks, it is important to show clearly what you are doing. Do this a step at a time.

> **Mark scheme**
>
> 1 method mark for $1.9 + 1.9 + 2.2 = 6$
>
> 1 method mark for 6×5.50
>
> 1 mark for £33.00.

Example: ABCDE is a pentagon.

Find an expression in terms of x and y for the perimeter of this pentagon.

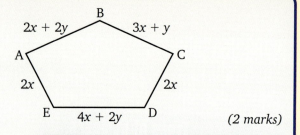

(2 marks)

Solution: For the first part of this question you have to form an expression. Then you need to simplify it.

$2x + 2y + 3x + y + 2x + 4x + 2y + 2x = 13x + 5y$

Mark scheme

1 mark is given for writing the expression.

1 mark is given for simplifying the expression.

Example: Mary's lucky number is x.

a Tom's lucky number is three less than Mary's lucky number.

Write down an expression for Tom's lucky number in terms of x. (1 mark)

b Jane's lucky number is four times Tom's lucky number.

Write down an expression for Jane's lucky number in terms of x. (1 mark)

c Jane's lucky number equals Mary's lucky number.

Write down and solve an equation to work out the value of x. (4 marks)

Solution: Parts **a** and **b** are routine questions about using a letter symbol to write algebraic expressions.

Part **c** is about using your answers to parts **a** and **b** to write down and solve an equation.

a Three less than Mary's lucky number is three less than x.

Three less than x is written as $x - 3$

So the expression for Tom's lucky number is $x - 3$

Study tip

A common incorrect answer to part **b** is $x - 3 \times 4$

This scores **no marks**.

Take care to include the bracket when addition or subtraction is done before multiplication.

b Four times Tom's lucky number is $x - 3$ multiplied by four.

When writing this as an expression the $x - 3$ must be put inside a bracket.

This makes sure that **all** of Tom's lucky number is multiplied by four.

So the expression for Jane's lucky number is $(x - 3) \times 4$

This is written better as $4(x - 3)$

Study tip

If you choose to use trial and improvement to work out x in part **c** you will score **no marks** because you are not following the instruction to 'Write down and solve an equation.'

c Jane's lucky number equals Mary's lucky number, so the equation is:

$4(x - 3) = x$

The next step is to solve your equation.

$4x - 12 = x$

$3x = 12$

$x = 4$

Study tip

Always check that your solution of an equation works. You can check your answer by doing the following.

Jane's lucky number $= 4(x - 3) = 4 \times (4 - 3)$
$= 4 \times 1 = 4$

Mary's lucky number $= x = 4$

So when $x = 4$, Jane's lucky number equals Mary's lucky number.

Mark scheme

1 mark for writing $x - 3$ 1 mark for writing either $(x - 3) \times 4$ or $4(x - 3)$ or equivalent.

1 mark for writing an equation correctly based on your answers to parts **a** and **b**. So, for example, if you had forgotten the bracket and given Jane's lucky number as $x - 3 \times 4$ you score 1 mark for writing the equation $x - 3 \times 4 = x$

1 mark for removing the bracket correctly.   1 mark for correct rearrangement of the equation.

1 mark for the correct answer.

Questions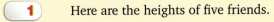

1   Here are the heights of five friends.

Alice 1.56 m   Ben 1.53 m   Chloe 1.61 m   Dev 1.62 m   Elly 1.55 m

The friends stand in a queue from tallest to shortest, with the tallest at the back.

  a   Who is at the back of the queue?

  b   Who is at the front of the queue?

  c   Who is in the middle of the queue?

2   Grey squares are used to make patterns.

Pattern 1   Pattern 2   Pattern 3

  a   Draw Pattern 4.

  b   How many grey squares are used to make Pattern 5?

  c   How many grey squares must be added to Pattern 5 if you wanted to make Pattern 8?

  d   How many grey squares are needed to make Pattern 20?

3   You can make a new number by swapping the positions of two digits in a number.

For example

You are given the number 23 187.

Swap 3 and 8 so that 23 187 → 28 137.

Swap 2 and 7 so that 28 137 → 78 132.

So 23 187 can be changed to 78 132 with two swaps.

  a   You are given the number 25 147.

    i   What is the largest number you can make using **one** swap?

    ii   What is the smallest number you can make using **two** swaps?

  b   You are given the number 3465.

    i   The 4 and 3 are swapped.
          Work out the difference between the old number and the new number.

    ii   The 5 and the 4 are swapped and then the 4 and the 6 are swapped.
          Work out the difference between the old and the new numbers.

  c   How many swaps does it take to change the number 1234 to the largest possible **even** number?
        Show the numbers you make after each swap.

4   Laura's friend has marked her homework for her but made some mistakes.
      Which parts has she marked wrongly?

  a $3^2 = 6$ ✓   **d** $\sqrt[3]{125} = 25$ ✗

  b $\sqrt{121} = 11$ ✓   **e** $4^3 = 12$ ✗

  c $64 = 4$ ✗   **f** $8^2 = 56$ ✓

F

5   Mary goes shopping in a local supermarket.

She buys:
3.5 kg of bananas at £1.20 per kilogram
three pens at £3.75 each
two notebooks at £2.20 each.

She pays with a £20 note.

How much change should she get?

G
F
E

6   Tim, Stacey and Afzal think of a number and give each other clues about it.

a   Here are the clues for Tim's number.

CLUE 1 My number is a multiple of 2.
CLUE 2 My number is a multiple of 5.
CLUE 3 My number is a three-digit square number.

Use the clues to work out the number Tim is thinking of.

b   Here are the clues for Stacey's number.

CLUE 1 My number is a two-digit factor of 48.
CLUE 2 When I add the digits of my number I do **not** get a factor of 48.

Use the clues to work out the number Stacey is thinking of.

c   Here is the clue for Afzal's number.

CLUE 1 My number is a two-digit square number that is also a cube number.

Use the clue to work out the number Afzal is thinking of.

E

7   Amy and Bill take part in a quiz. The rules are:

For each correct answer score +5 points.
For each wrong answer score −3 points.

There are five questions in each round.

a   After the first round Amy has three questions
correct and Bill has four questions correct.
What are their scores?

b   In the second round Bill scores +1. This is added on to his previous score.

Amy wants to have more points than Bill.

What is the least number of questions that Amy must get right in the second
round, so that she has more points?

8   James has 30 marbles.

$\frac{1}{2}$ of the marbles are red.

$\frac{1}{3}$ of the marbles are blue.

The rest of the marbles are green.

Work out the fraction of the marbles that are green.

Give your answer in its simplest form.

9 Shelving units can be bought in the widths shown in the table.

Shelves	Type A	Type B	Type C
Width (metres)	0.65	0.85	1.05
Cost	£37.99	£47.99	£57.99

a The diagram shows two Type A shelves, two Type B shelves and three Type C shelves placed next to each other along a wall.

A	A	B	B	C	C	C

i What is the total length of the shelving units?

ii What is the total cost of the shelving units?

b In a school there is a classroom with one wall of length 6.8 metres.

The school wants to build shelving units along this wall, using as much of the wall as possible.

What shelving units should the school buy to do this?

You **must** show working to justify your answer.

10 Lisa uses 60% of a tin of paint to paint a fence.

What is the smallest number of tins of paint she needs to buy to paint eight fences?

You **must** show your working.

11 The points A, B, C, D and E are plotted on the grid.

a Which two points lie on the line $x = 4$?

b Which two points lie on the line $y = 7$?

c Which two points lie on the line $y = x$?

d Which two points lie on the line $x + y = 11$?

e Which two points lie on the line $y = 2x - 1$?

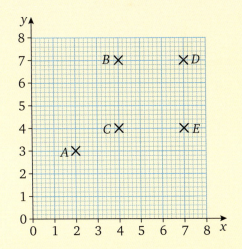

12 **a** Solve the following equations.

i $2x + 11 = 3$

ii $4(y - 2) = 22$

iii $7z + 1 = 10 - 3z$

b In an election, 200 people voted for three candidates.

Sarah wins the election.

She beats Tom by 27 votes and Dilip by 43 votes.

How many votes does Sarah get?

D

13 The Ancient Egyptians used unit fractions like this:

$$\widehat{\text{III}} = \frac{1}{3}$$

A unit fraction has a numerator of 1.

The Ancient Egyptians made other fractions by adding unit fractions together.

For example, $\frac{2}{3}$ can be made by adding $\frac{1}{2}$ and $\frac{1}{6}$

$\frac{1}{2} + \frac{1}{6} = \frac{3}{6} + \frac{1}{6} = \frac{4}{6} = \frac{2}{3}$

> **Hint**
>
> You can use either any pair of these fractions or all three.

a What fractions can you make by adding $\frac{1}{2}$, $\frac{1}{3}$, and $\frac{1}{4}$?

b An Ancient Egyptian farmer shares five loaves between eight people working in his fields.

   i What fraction of a loaf does each worker get? Give your answer as two unit fractions added together.

   ii Describe how the farmer cuts the loaves to make sure each worker gets an equal share.

c Repeat part **b** for three loaves shared between four workers.

14 The distances in metres around the outside of a field are shown on the diagram. *BD* is a path across the field.

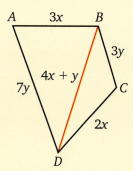

a Adrian walks round the outside of the field from *A* back to *A* again.

Show that the distance that Adrian walks can be written as $5(x + 2y)$ metres.

b Beth walks from *A* to *B*, then along the path *BD* and back along *DA* to *A*.

Write down a simplified expression for how much further Adrian walks than Beth.

c Beth says $y > x$. Is she right? Give a reason for your answer.

15 The *n*th term of sequence **A** is $2n + 3$

The *n*th term of sequence **B** is $4n - 1$

a Work out the difference between the 20th term of sequence **A** and the 20th term of sequence **B**.

b Show that 11 is a term in both sequence **A** and sequence **B**.

16 A painter needs 15 litres of orange paint.

He decides to make the orange paint by mixing red paint and yellow paint in the ratio $3 : 2$

He measures the amount of paint he needs using 200 ml tins.

How many 200 ml tins of each colour of paint are needed?

> **Hint**
>
> 1 litre = 1000 ml

17 **a** Express 144 as the product of its prime factors. Write your answer in index form.

b Find the highest common factor (HCF) of 30 and 144.

c Find the least common multiple (LCM) of 60 and 144.

18 Dave and Debbie want to go out.

They can go to either the cinema, an exhibition or a show at a theatre.

If they go to these they will need to travel by taxi each way.

They have a choice of two taxi companies.

Use this information to work out which would be the cheapest option for them.

	Distance from home	Ticket price per person
Cinema	8 miles	£8.50
Exhibition	6 miles	£10
Show	5 miles	£15.50

	Charge per trip
A2B Cabs	£2.60 per mile
Sapphire Taxis	£1.20 per mile plus £10

19 Ben sees this entry in an online catalogue.

In addition to the special offer, there is a choice of payment options.

LAPTOP

Ram	3 GB 667 MHz
Hard Drive	320 GB 5400 rpm
Processor	Core 2 Duo 2.26 GHz
Screen	15.4 Inch TFT

£540.00

SPECIAL OFFER
20% OFF MARKED PRICE

		Terms
Option A	Cash payment	10% discount
Option B	6 months to pay	5% deposit Pay the remaining amount monthly in six equal instalments

Ben buys the laptop and decides to pay using **Option B**.

Give full details of what Ben has to pay.

20 Which of these numbers is the smallest?

$a = 4 \times 4^3 \times 4^5$ $b = \dfrac{4^{10}}{4^2}$ $c = \dfrac{4^3 \times 4^6}{4^2}$

You must show all your working.

Bump up your grade

To get a Grade C you must be able to use index notation.

21 A box of sweets falls on the floor.

When the sweets are put together in groups of three, there is one left over.

When the sweets are put together in groups of four, there is one left over.

When the sweets are put together in groups of five, there is one left over.

What is the smallest possible number of sweets that are in the box?

22 **a** Solve the inequality $3x - 5 > 16$

b n is an integer and $-8 \leqslant 2n < 6$

Work out all the possible values for n

C

Examination-style questions

1 **a** Giant tubes of fruit gums cost £1.50.
How many giant tubes can Helen buy for £10? *(2 marks)*

 b Carl spends £2.43 on sweets. He pays with one £10 note.
 i How much change does Carl receive? *(1 mark)*
 ii This change is given using the smallest number of notes and coins possible.
How is the change given? *(2 marks)*

AQA 2005

2 **a** A sequence starts 2, 7, 17, ……
The rule for finding the next term in this sequence is to multiply the previous term
by 2 and then add on 3
Work out the next term. *(1 mark)*

 b The rule for finding the next term in a different sequence is to multiply the previous
term by 2 and then add on a, where a is an integer.
The first term is 8 and the fourth term is 127

8 …… …… 127
Work out the value of a. *(4 marks)*

AQA 2009

3 The graph shows Ben's progress on a sponsored walk.

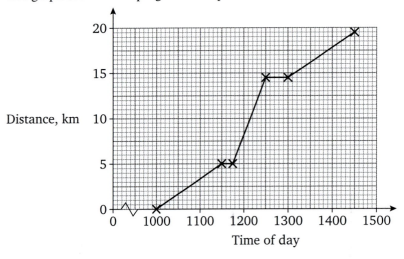

 a **i** How long is the walk? *(1 mark)*
 ii Work out the total time that Ben stops during his walk. *(1 mark)*
 iii Between which times does Ben walk the fastest? Explain your answer. *(2 marks)*

 b Sally walks with Ben until they stop at 12:30.
She stops for half an hour longer than Ben but then walks twice as fast as he does.
At what time does Sally catch Ben up?
You **must** show your working. *(3 marks)*

AQA 2004

4 Alan has some unknown weights labelled a and b and some 5 kg and some 10 kg weights.
He finds that the following combinations of weights balance.

 a Find the value of a. *(1 mark)*

 b Find the value of b. *(2 marks)*

 c Alan also has some unknown
weights labelled c.

 He finds that $5c + 2b = c + 6a$
Work out the value of c.
(4 marks)
AQA, 2007

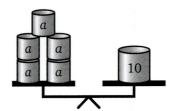

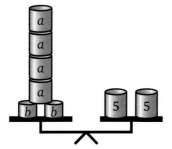

Section 3

Before attempting this chapter, you will need to have covered the following topics:

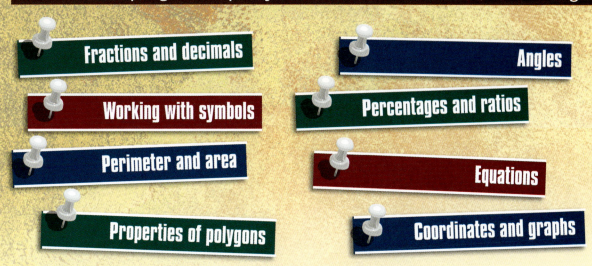

Fractions and decimals

Angles

Working with symbols

Percentages and ratios

Perimeter and area

Equations

Properties of polygons

Coordinates and graphs

All these topics will be tested in this chapter and you will find a mixture of problem solving and functional questions. You won't always be told which bit of maths to use or what type a question is, so you will have to decide on the best method, just like in your exam.

Example: This is a regular hexagon.

It has a perimeter of 18 centimetres.

Not drawn accurately

Three of these hexagons are used to make this shape.

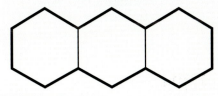

> **Hint**
>
> The two sides inside the shape are not part of the perimeter.

Not drawn accurately

Work out the perimeter of the shape.   (3 marks)

Solution: As a diagram is given, you can count the number of sides.

The number of sides = 6

Each side of the regular hexagon = 18 cm ÷ 6

= 3 cm

The shape has 14 sides.

Perimeter of shape = 14 × 3 cm

= 42 cm

> **Study tip**
>
> Mark each side of the diagram as you count so that you get the total correct.

Mark scheme

- 1 mark for attempting 18 ÷ 3 or obtaining the side length as 3 cm.
- 1 mark for multiplying 14 by the side length.
- 1 mark for the correct final answer of 42 cm.

> **Study tip**
>
> You are allowed a calculator in Unit 3 so use it to work out 14 × 3

Example: The diagram shows a right-angled triangle.

a Work out the value of *x*.

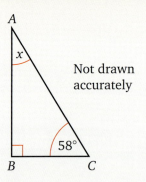

Not drawn accurately

(2 marks)

b *X* is a point on *AC* so that *BX* = *CX*

The triangle is cut along the line *BX* to make two triangles.

Show that triangle *ABX* is an isosceles triangle.

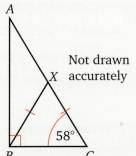

Not drawn accurately

(2 marks)

Solution: **a** Start by thinking about what you know about the angles of a triangle.
The angles of a triangle add up to 180°.
A right angle is 90°.

So the angle at *A* must be   *180° − 90° − 58° = 32°*

x = 32°

b *BX* = *CX*. Think about what this tells you about the triangle *BXC*.
This means **triangle *BXC* is isosceles.**
Think of what you know about isosceles triangles.
Two sides are the same length and the base angles are the same.

So **angle *XBC* must also be 58°.**

> **Study tip**
>
> When a question says 'Show that...', you must make sure that you set your working out very clearly so that you do not miss out any steps in your working.

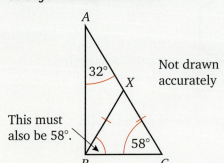

This must also be 58°.

Not drawn accurately

> **Study tip**
>
> It can help to mark the angles on the diagram.

Because angle *B* is 90°, angle *ABX* must be **90° − 58° = 32°**

Triangle *ABX* has two angles which are the same, at *A* and *B*.

So this triangle must also be isosceles.

> **Mark scheme**
>
> - 1 mark for the method of working out angle *A*.
> - 1 mark for the final answer 32°.
> - 1 mark for seeing that angle *XBC* is 58°.
> - 1 mark for working out angle *ABX* as 32° and explaining why this means the triangle is isosceles.

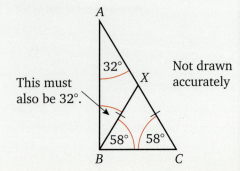

This must also be 32°.

Not drawn accurately

Questions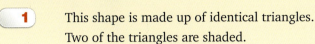

1 This shape is made up of identical triangles.
Two of the triangles are shaded.

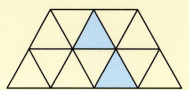

    a How many more triangles must be shaded so that $\frac{3}{4}$ of the shape is shaded?

    b Which of these percentages is the same as $\frac{3}{4}$?

        25%   34%   43%   75%

2 Choose a word from the list that describes each of these angles.

    acute   obtuse   reflex   right

a    **b**   **c**   **d**

3 Margaret goes to a shop to buy sandwiches for her friends at work.

This is her list of what to buy.

2 ham and cheese sandwiches

5 chicken sandwiches

4 bags of crisps

chicken: £2.85

ham and cheese: £2.65

crisps: 40p

She has a £20 note in her purse.
Will she have enough money?
You **must** show all your working.

4 Name each shape. Pick from the list given each time.

    a

    rectangle   hexagon   trapezium   pentagon   square

    b

    oblong   parallelogram   trapezium   square   rectangle

5 **a** Copy the number line. Fill in the two empty boxes with the numbers represented on the number line.

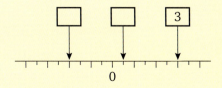

    b a, b, and c are three different numbers.

       $a \times b$ is negative.

       $b \times c$ is positive.

       Which of the following statements, A, B and C, is correct?

       A: $a \times c$ is positive.

       B: $a \times c$ is negative.

       C: $a \times c$ may be positive or negative.

       You **must** give a reason for your answer.

G
F

F

6   Joss and Hayley are finding points on this grid.

Joss draws a cross on points where $x + y = 6$

Hayley draws a circle around points where $y - x = 2$

Here is the grid after they have each had two goes.

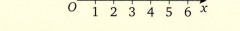

a   If they continue finding points, how many points on this grid will have a circle drawn around them?

b   Write down the coordinates of the point that Joss could plot that is on the x-axis.

c   Write down the coordinates of the point that Hayley could plot that is on the y-axis.

d   How many points will have a cross with a circle drawn around it?

Write down the coordinates of these points.

7   Shapes A and B are drawn on a square grid.

The area of shape A is 20 cm^2.

Work out the area of shape B.

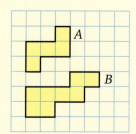

8   In the diagram:

a is the smallest angle

b is 10° bigger than a

c is 20° bigger than b

d is 30° bigger than c.

Work out the size of angle a.

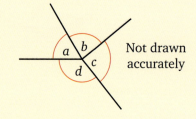

Not drawn accurately

9   The perimeter of a regular pentagon is 64 cm.

Work out the length of each side.

10   **a**   Davinda buys 40 carpet tiles.

One-fifth of the tiles are blue and the rest are white.

Blue tiles cost £4.15 each.

White tiles cost £2.75 each.

Work out the total cost of the tiles.

b   Each tile is a square of side 30 cm.

Davinda arranges the 40 tiles in a 10 by 4 rectangle on the floor.

Work out the perimeter of the rectangle that he has made.

11   Jack is 1.2 kg heavier than Kylie.

Together they weigh 18 kg.

Work out Jack's weight.

12   A wall in a room is 330 centimetres wide.

Su Ling wants to hang two pictures on the wall.

She wants the distances between the pictures and the ends of the wall to be the same.

Each picture is 45 centimetres wide.

Work out the distance, d, between the pictures.

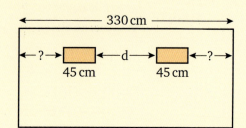

13 **a** Simplify fully $\dfrac{14d}{7}$

    b $s = -5$ and $u = 4$
       Work out the value of $2s - 3u$

    c Simplify $7g + 4f - g - 6f$

14 The diagram shows the plan for a new park.

The local council says that the percentage of the area allocated to each purpose is:

    gardens 52%

    children's play area 4%

    sports pitches 34%

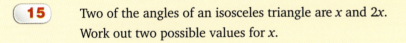

    a Erin says 'There is a mistake with these figures'.
       Give a reason why Erin knows this.

    b The area of the park is 0.675 km².

      i The local council says the figure for the area of the gardens should have been 62%.
         What is the area for the gardens?
         Give your answer to three decimal places.

      ii In a change to the plan, 0.054 km² of the park is turned into a car park.
         What percentage of the park is this?

Not drawn accurately

15 Two of the angles of an isosceles triangle are x and $2x$.
    Work out two possible values for x.

16 **a** Your family are going on holiday to Germany.
       You change £2500 into euro before you leave.
       The exchange rate is £1 = 1.142 euro
       How many euro do you receive?

    b During your holiday you change another £360 into euro.
       The exchange rate is now 1 euro = £0.90
       How many euro do you receive?

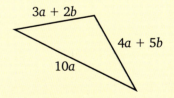

    c At the end of your holiday you want to change 150 euros into pounds.
       The exchange rate in Germany is £1 = 1.165 euro
       The exchange rate in Britain is £1 = 1.172 euro
       In which country will you get more pounds for your 150 euro?
       Give a reason for your answer.

17 Here is a triangle.

    a Write down an expression for the perimeter of the triangle.
       Simplify your answer.

    b When $b = 3$, the perimeter of the triangle is 55 cm.
       Work out the value of a.

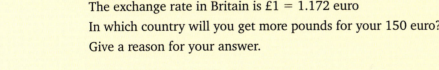

18 Work out the angles marked with a letter.

    a

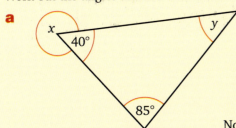

    b

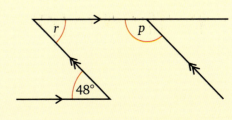

Not drawn accurately

F E

F E D

E

F E D

E
D

19   **a**   Round these numbers to one significant figure.

   i   8.256

   ii   19.2

   iii   7821

   b   Use your answers to part **a** to estimate $\dfrac{19.2 \times 7821}{8.256}$

20   Here are four sketch graphs.

Graph A   Graph B   Graph C   Graph D

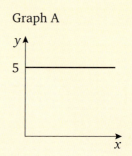   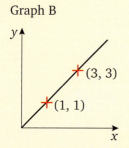   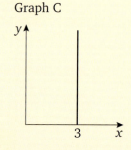   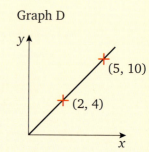

Pick the equation from those given below that each graph represents.

$y = 3$ $y = 2$ $y = x$ $x = 3$

$y = 5$ $x = 5$ $y = \frac{1}{2}x$ $y = 2x$

D

21   The diagram shows the cross-section of a roof.

The two sides of the roof are perpendicular.

One side of the roof slopes at 35° to the horizontal.

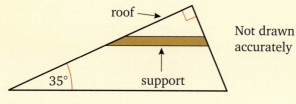

Not drawn accurately

A horizontal support is put in.

The support is made from a rectangular piece of wood.

To make it fit, the wood is cut along the dotted lines.

Not drawn accurately

Work out the values of x and y.

22   The boat *Clarabel* breaks down and sends out a distress call.

The call is heard by two other boats, *Aramis* and *Bellamy*.

They set out to intercept *Clarabel*.

Aramis sets out on a bearing of 055°.

Bellamy sets out on a bearing of 260°.

Copy the diagram and mark the position of *Clarabel*.

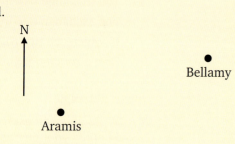

23   Which of these shapes has the largest area?

Circle   Triangle

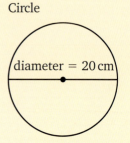   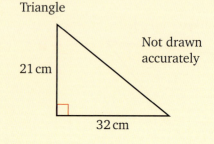

Not drawn accurately

You **must** show your working.

24   Ahmed has the following test results.

French: 61%

ICT: 75 out of 120

In which subject did he get the better score?

You **must** show your working.

25   Alice is four years younger than Ben.

She is nine years older than Carly.

The total age of all three people is 79.

How old is Alice?

26   A rectangle has a length that is twice the width.

Four of these rectangles make this shape.

Work out the perimeter of the shape.

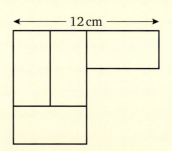

D

27   An activity centre organises climbing trips for the public.

a   Each member of staff at the centre must not take more than 8 members of the public in their group.

A party of 76 people book a trip.

12 members of staff are available to take the party.

i   Are there enough members of staff to take all the people? You **must** show all your working.

ii   What is the ratio of staff to people for this trip? Give your answer in the form $1 : n$

D C

b   In 2009 there were 18 726 people who went on climbing trips.

In 2010 this number increased to 19 871.

Work out the percentage increase.

Give your answer to one decimal place.

> **Bump up your grade**
>
> To get a Grade C you should be able to work out a percentage increase or decrease.

28   **a**   Simplify the following expression.

$6(x - 2) - 2(2x + 3)$

b   Two numbers in the following expression have been covered up.

$\bullet x - \bullet(4 - x)$

The expression simplifies to $5x - 8$.

Work out the two numbers.

C

29   Here is part of a number line.

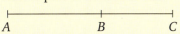

A B C

$AB : BC = 4 : 3$

If $A = 2.8$ and $C = 14.7$, work out the value of B.

C

30 The points P, Q and R are on a straight line.

$PQ : QR$ is $3 : 1$

Work out the coordinates of point Q.

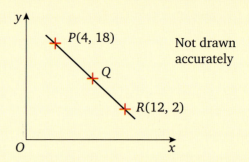

Not drawn accurately

31 An interior angle of a regular polygon A is $135°$.

Another regular polygon B has two sides fewer than polygon A.

Work out one of the interior angles of polygon B.

Examination-style questions k

1 **a** PQR is a straight line.

Find the value of x.

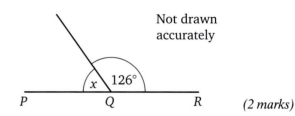

Not drawn accurately

(2 marks)

b The three lines shown below meet at a point.

Find the value of y.

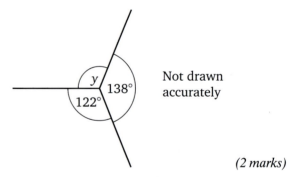

Not drawn accurately

(2 marks)

c In the diagram, AB is parallel to CD.

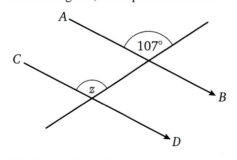

Not drawn accurately

Find the value of z.

(1 mark)

AQA 2006

2 Complete this table.

Expression	Value
$2x$	8
$5x$	
$2x + 3y$	5
y	
$3x - y$	

(2 marks)

AQA 2006

Section 4

Before attempting this chapter, you will need to have covered the following topics:

Fractions and decimals

Working with symbols

Perimeter and area

Properties of polygons

Reflections, rotations and translations

Area and volume

Trial and improvement

Construction

Quadratics

Angles

Percentages and ratios

Equations

Coordinates and graphs

Formulae

Measures

Enlargements

Loci

Pythagoras' theorem

All these topics will be tested in this chapter and you will find a mixture of problem solving and functional questions. You won't always be told which bit of maths to use or what type a question is, so you will have to decide on the best method, just like in your exam.

Example: The diagram shows part of a golf course. There are nine holes.

The 7th hole is at *P*.

The 8th hole is at *Q*.

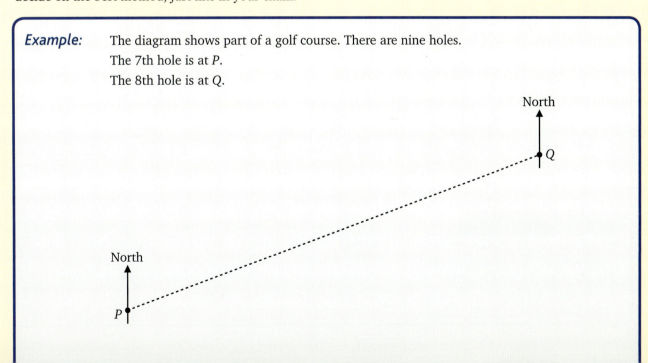

The diagram has been drawn using a scale of 1 : 200

a   What is the actual distance, in metres, between the 7th hole and the 8th hole?   *(3 marks)*

b   The 9th hole is 16 m from Q and on a bearing of 300° from Q.

Mark, with a cross, the position of the 9th hole on the diagram.   *(3 marks)*

Solution: **a**   The scale 1 : 200 means 1 cm on the diagram is equal to 200 cm on the actual golf course.

There are 100 cm in 1 metre.

200 ÷ 100 = 2

So 200 cm = 2 metres

> **Study tip**
>
> In exams, there is a tolerance of ± 2 mm on measurements. So, if your measurement is between 11.8 and 12.2 cm you can get full marks.
>
> However, always take care and check. Remember that measurements are **not** always a whole number of centimetres.

So the scale 1 : 200 also means 1 cm on the diagram is equal to 2 metres on the actual golf course.

Start by measuring the length PQ on the diagram.

PQ = 12 cm

So the actual distance between the 7th hole and the 8th hole = 12 × 2 = *24 metres*

> **Study tip**
>
> You will lose 1 mark if you do not give your answer in metres.

> **Mark scheme**
>
> • 1 mark for measuring PQ to get 11.8 cm to 12.2 cm
> • 1 mark for using the scale 11.8 to 12.2 × 200
> • 1 mark for dividing by 100; this mark can be earned in two ways. Either for changing the scale to 1 cm : 2 m or by working out the answer in centimetres and then dividing by 100.

b   Convert 16 m to centimetres using the scale.

16 ÷ 2 = 8 cm

Draw a line from Q to represent the bearing 300°.

This is the blue line on the diagram.

Mark a cross on this line 8 cm from Q.

This is the position of the 9th hole.

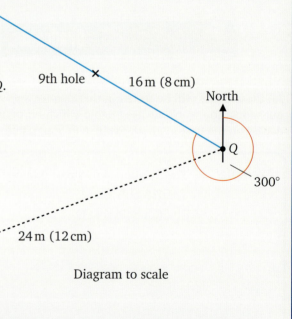

9th hole ✕   16 m (8 cm)

North

Q

300°

24 m (12 cm)

North

P

Diagram to scale

> **Study tip**
>
> In exams, you are allowed a tolerance of ± 2° in the accuracy of the angle and ± 2 mm in the accuracy of the length.

> **Mark scheme**
>
> • 1 mark for working out that the required length from Q is 8 cm.
> • 1 mark for a line showing the bearing, with an angle of 298° to 302° allowed.
> • 1 mark for marking a cross along this line 8 cm from Q, with a length of 7.8 cm to 8.2 cm allowed.

Example: A company makes toy building blocks in the shape of cubes.

There are three sizes.

Some blocks have sides of 2 cm, some 4 cm and some 8 cm.

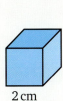 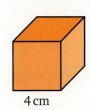 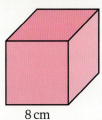

2 cm   4 cm   8 cm

The blocks are sold in boxes.

In a box there is:

- a bottom layer made up of the 8 cm blocks
- a middle layer made up of the 4 cm blocks
- a top layer made up of the 2 cm blocks.

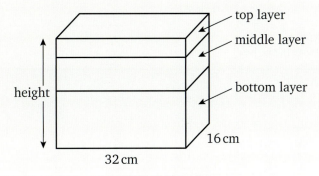

a   What is the height of the box?   *(1 mark)*

b   Work out how many blocks are in the box.   *(3 marks)*

Solution: **a**   Think of an 8 cm block with a 4 cm block on top and a 2 cm block on top of that.

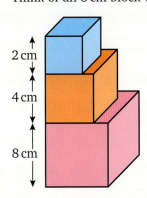

The height is   8 + 4 + 2 = 14 cm

This must be the height of the box.

The height of the box is 14 cm.

Study tip

Don't forget to read everything carefully and use the diagrams to make sense of the question.

Mark scheme

- 1 mark for working out the height of the box correctly.

b   Start by thinking of the bottom layer.
How many 8 cm blocks will there be?

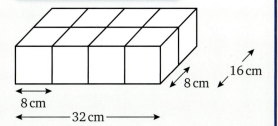

32 ÷ 8 = 4 means there are 4 blocks along the front

16 ÷ 8 = 2 means there are 2 rows

This will mean there are 4 x 2 = 8 of these blocks.

Repeat this working for the other layers.

For the 4 cm blocks in the middle layer.

32 ÷ 4 = 8 and 16 ÷ 4 = 4

So there are 8 of these blocks along the front and 4 rows.

There are 8 x 4 = 32 of these blocks.

For the 2 cm blocks in the top layer.

32 ÷ 2 = 16 and 16 ÷ 2 = 8

So there are 16 of these blocks along the front and 8 rows.

There are 16 x 8 = 128 of these blocks.

Now add up all the blocks.

Altogether there are 8 + 32 + 128 blocks

= 168

There are 168 blocks in the box.

> ### Mark scheme
>
> - 1 mark for knowing how to work out the number of blocks in one layer
> - 1 mark for working out the blocks in each of the three layers
> - 1 mark for the final correct answer

Questions

G

1   Stephen is making shapes using centimetre cubes.

    a   What is the volume of this shape?

1 cm

    b   Stephen has eight cubes altogether.

        He decides to make a large cube which is 3 cm high.

        How many more centimetre cubes does he need
        to complete his large cube?

2   A flight from Jersey to Southampton Airport on Easyfly Airways lasts 50 minutes.
A flight leaves Jersey at 09:40.

    a   What time will it arrive at Southampton Airport?

    b   Mr Bell has a meeting in Southampton.
        The meeting starts at 11:00.
        It takes 45 minutes for Mr Bell to get from the airport to the meeting.
        Does Mr Bell arrive on time for the meeting?
        Give a reason for your answer.

    c   The flight back to Jersey leaves Southampton at 17:30.
        Mr Bell must be at the airport at least 1 hour before the flight leaves.
        He allows 45 minutes to get from the meeting to the airport.
        What is the latest time Mr Bell must leave the meeting?

3 **a** Copy the diagram and add two more lines to it to make a polygon.
The dotted line must be a line of symmetry of your polygon.

b Here are the names of some polygons.

hexagon   kite   octagon   pentagon   rhombus

Write down the one which is the name of the shape you made in part **a**.

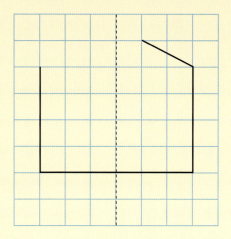

4 **a** Draw a circle with radius 5 centimetres.

b On your circle draw a chord of length 6 centimetres.

c Draw a diameter of your circle that is perpendicular to your chord.

5 The diagram shows a regular hexagon with all the possible diagonals.

a On a copy of the diagram shade all the triangles that are congruent to triangle *A*.

b How many triangles, congruent to triangle *B*, are in the diagram?

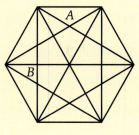

6 The point *A* is plotted on this grid.

a Write down the coordinates of *A*.

b The point *B* has coordinates (7, 6).
The point *C* has coordinates (3, 5).

  i Measure the length of the line *AB*.

  ii Measure angle *CAB*.

c Explain why triangle *ABC* is isosceles.

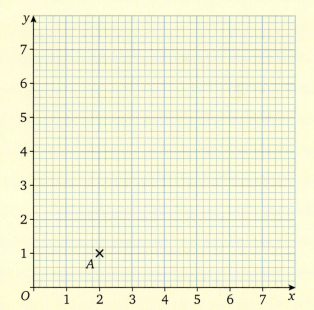

7 The diagram shows a sketch of a ladder against a wall.

a Draw the diagram accurately using a scale of 1 centimetre to 1 metre.

b Use your diagram to work out the length of the ladder. *ℓ*

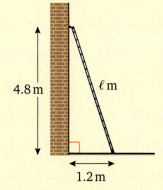

F

8 **a** This symbol is the emblem of the Isle of Man.

Write down the order of rotational symmetry of the symbol.

b Shade four more squares so that this shape has
rotational symmetry of order 4 about its centre, *C*.

9 Clark is driving his car in France. His journey is 320 kilometres.

a Clark says, '320 kilometres is not as far as 320 miles.'

Is Clark correct?

You must give a reason for your answer.

b A milometer records the number of miles the car has travelled.

At the start of the journey the milometer reads

What will the milometer read at the end of the journey?

10 A car hire company works out the cost of hiring a car using the following formula:

cost of hire = fixed charge + number of days × daily rate

	Fixed charge	Daily rate
Small car	£10	£25
Medium size car	£15	£28
Large family car	£18	£30

a Linda wants to hire a car for 3 days.

How much extra will it cost her to hire a medium size car than a small car?

b Mr Brown hired a large family car.

He was charged £178.

Mr Brown complained that he was not charged the correct amount.

Showing all your working, explain why the company must have made a mistake.

11 *AB* and *CD* are straight lines.

AD and *DC* are perpendicular.

Work out the value of angles *x*, *y* and *z*.

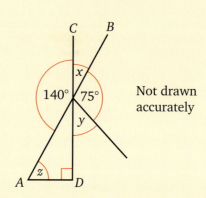

Not drawn
accurately

12   One-half of a number is 15.

What is one-third of the number?

13   Two rectangles *A* and *B* can fit together to make a square.

The height of *A* is the same as the width of *B*.

The width of *A* is 3 cm.

The perimeter of *A* is 26 cm.

What is the perimeter of *B*?

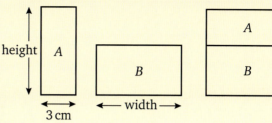
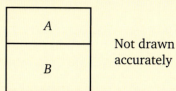

Not drawn accurately

14   **a**   Reflect this shape in the line $x = 3$

b   The whole shape is also symmetrical about a different line. Write down the equation of this line.

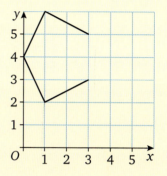

15   All of these shapes are drawn accurately.

Show clearly that each shape has the same area.

16   There are some red counters, white counters and blue counters in a bag.

The total number of counters is 30.

$\frac{1}{3}$ of the counters are red.

$\frac{1}{2}$ of the counters are white.

Work out the fraction of the counters that are blue.

Give your answer in simplest form.

E

17

a Complete the table below for the graph $y = 2x - 3$

x	−1	0	1	2	3
y		−3			3

b Draw the graph $y = 2x - 3$

c Explain how the graph can be used to solve the equation $2x - 3 = 0$

18 Karla has hidden Jeremy's phone and mp3 player in a field.

She gives him these instructions to find his phone:

> From the gate, move 10 metres north.
>
> Then move 8 metres east.
>
> Then move 2 metres south.
>
> Your phone is here.

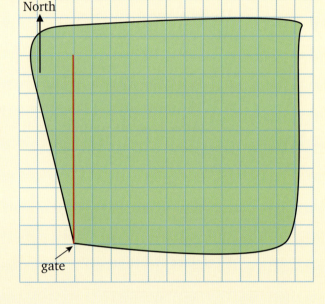

North

gate

a The first step of the instructions has been drawn on the diagram.

Copy and complete the diagram to show all the steps and mark the point where the phone is.

b Karla says:

> 'From where I hid the mp3 player I moved 3 metres south and 9 metres west to get back to the gate.'

Copy and complete the following two instructions that will take Jeremy from where the phone is to where the mp3 player is.

Move _____ metres _____

Then move _____ metres _____

19 The graph shows the path of a cricket ball when it is thrown from a fielder to the wicketkeeper.

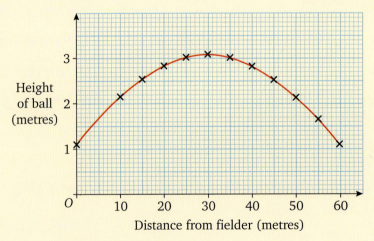

Height of ball (metres)

Distance from fielder (metres)

a How far is the wicketkeeper from the fielder?

b At what height does the wicketkeeper catch the ball?

c What is the greatest height above the ground that the fielder throws the ball?

20 **a** Rachel draws two different quadrilaterals each containing **two** right angles.

    i Her first quadrilateral has **one** line of symmetry.
Draw **and** name this quadrilateral.

    ii The second quadrilateral that Rachel draws has **no** lines of symmetry.
Draw **and** name this quadrilateral.

  b Rachel tries to draw a quadrilateral with exactly three right angles.
Explain why she finds this impossible.

21 A hexagon with equal sides is cut along the dotted lines.
The dotted lines are parallel to the top and bottom sides of the hexagon.
The four pieces A, B, C and D all have the same height.

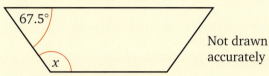

Not drawn accurately

  a What is the mathematical name of each of the four pieces?

  b Draw a diagram to show how the four pieces can be put together to form a parallelogram.

  c Shape D has an angle of 67.5° as shown below.

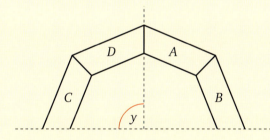

Not drawn accurately

Work out the value of x.

  d The four pieces can be arranged symmetrically as shown below.
The dotted lines are straight.

Not drawn accurately

Explain why angle y must be a right angle.

22 Simplify fully the following expressions.

  a $x + x + 2$ **c** $x + x \times 2$

  b $x \times x \times 2$ **d** $x \times x + 2$

23

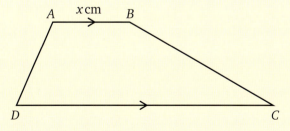

Not drawn accurately

$AB = x$ cm
BC is three times longer than AB.
CD is 2 cm longer than BC.

  a Write down an expression in terms of x for the length CD.

  b The perimeter of the shape is $2(4x + 3)$ cm.
Work out the length of AD in terms of x.

E

G
E
C

E
D

D
C

D
C

24 In a test there are 10 questions.

If you attempt a question and get it right you score 5 marks.

If you attempt a question and get it wrong you lose 2 marks.

If you do **not** attempt a question you lose 10 marks.

a   The table shows how three friends, Andrew, Bill and Clare, do in the test.

Name	Number of questions attempted	Number of questions correct
Andrew	10	6
Bill	9	8
Clare	8	7

In the test the mark for a Merit is 30.

The mark for a Pass is 20.

How well do the three friends do in the test?

You must show working to justify your answer.

b   Tim takes the test.

He scores −7.

How many questions did Tim not attempt?

You must show working to justify your answer.

D

25 Here are details of Sudhir's bicycle journey.

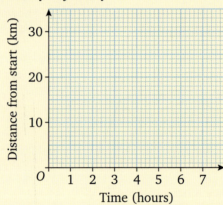

Stage 1:   After the start he cycles at a speed of 12 km/h for $2\frac{1}{2}$ hours.

Stage 2:   He stops for 30 minutes.

Stage 3:   He cycles back towards the start for 1 hour travelling 10 km.

Stage 4:   He stops for another 30 minutes.

Stage 5:   He cycles back to the start at a speed of 8 km/h.

On a copy of this grid draw a distance–time graph to represent Sudhir's journey.

26 The diagram shows an irregular hexagon drawn on a centimetre square grid.

The dotted line shows an enlargement of one of the sides of the hexagon.

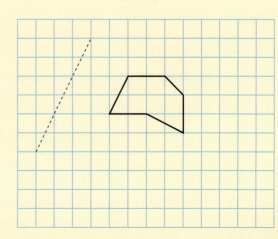

a   Copy the diagram and complete the enlargement.

b   Mark the centre of enlargement on the grid.

c   Write down the scale factor of the enlargement.

27 This solid shape is made from twelve cubes.
It is drawn on an isometric grid.
Draw three copies of this centimetre grid.

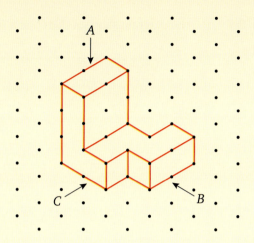

On your copies of the grid:

a draw the plan view from *A*

b draw the front elevation from *B*

c draw the side elevation from *C*.

28 Write down two different transformations which map *A* onto *B*.

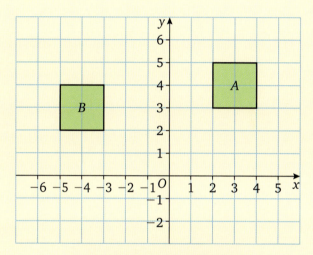

29 The volume of this cuboid is 480 cm³.

a Work out the width of the cuboid.

b The shape is cut in two, down the dotted lines, and then glued together to make a new cuboid.

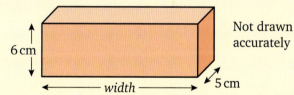

Not drawn accurately

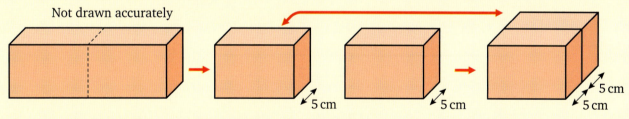

The surface area of the original cuboid is 412 cm².
Melissa says that the surface area of the new cuboid is the same as the original one.
Is Melissa correct?
You must show all your working to justify your answer.

30 The price of tickets for a boat trip to the Farne Islands is:

Adults   £12   Children   £8

x adults and *y* children go on a boat trip to the Farne Islands.
£*T* is the total price of their tickets.

a Write a formula for *T* in terms of *x* and *y*.

b On one boat trip the total price of the tickets is £672.
The number of children's tickets sold is 18.
How many adult tickets are sold on this trip?

D

31 The diagram shows a sketch of a triangular field, *ABC*.

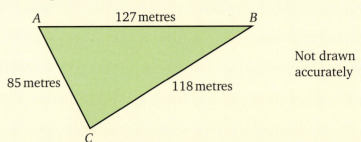

Not drawn accurately

a Using ruler and compasses only, construct an accurate scale drawing of the triangle.
Use a scale of 1 cm to 10 metres.

b In the field there is a telegraph pole.
The telegraph pole is 44 metres from *A* and 63 metres from *C*.
How far is the telegraph pole from *B*?

32 The diagram shows a centimetre grid with the points *A*, *B*, *C* and *D* marked.

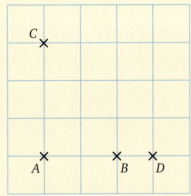

On a copy of the grid draw:

a the locus of the point that is an equal distance from *A* and *B*

b the locus of the point that is an equal distance from *A* and *C*

c the locus of the point that is an equal distance from *C* and *D*.

D
C

33 **a** Copy and complete the table for $y = x^2 + 5$

x	−3	−2	−1	0	1	2	3
y							

b Draw the graph of $y = x^2 + 5$ for values of *x* from −3 to +3.

C

34 Plot the points *X*(1, 1), *Y*(10, 2) and *Z*(6, 9) on a centimetre grid.

Using a ruler and compasses construct:

a the perpendicular bisector of the line *XY*

b the bisector of angle *XYZ*.

35 A rectangle has length x^2 and width $x + 3$.

The area of the rectangle is 40 cm².

Use trial and improvement to work out the value of *x*.

Give your answer to one decimal place.

36   Mrs Sim has this photo frame.

The photo frame is designed for a photograph of height 24 cm and width 17.5 cm.

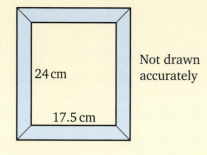

24 cm

17.5 cm

Not drawn accurately

Mrs Sim has a photograph of her daughter on her digital camera.

The print of this photograph has height 8 cm and width 5 cm.

8 cm   Not drawn accurately

5 cm

She wants to print an enlargement of this photograph so that it fits the photo frame exactly.
Is this possible?
You must show working to justify your answer.

37

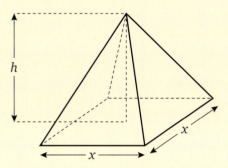

Not drawn accurately

h

x

x

The volume V of this square-based pyramid is given by the formula $V = \frac{1}{3}x^2h$ where x and h are measured in centimetres.

a   Work out the volume of the pyramid when $x = 6$ and $h = 8$

b   Aaron has been asked to make a square-based pyramid.

The height must be 2 cm bigger than the length of the base.

The volume of the pyramid must be 125 cm³.

He uses trial and improvement to work out the length of the base, x.

Complete his worksheet to find x to one decimal place.

Length of base, x cm	Height, h	x^2	Volume, cm³	
7	9	49	147	too big

38   A train usually travels along a section of track at 90 km/h for 6 minutes.

During repairs, the speed limit along this section is reduced to 20 km/h.

How much time will this add on to the journey?

39   Work out the length x in this triangle.

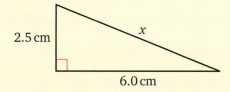

2.5 cm

x

6.0 cm

C

40 A machine cuts circular discs of diameter 5 cm from a sheet of rectangular plastic.

The dimensions of the sheet of plastic are 1.2 m by 0.8 m.

The machine leaves the following horizontal and vertical gaps:

- 2 mm between each disc
- 2 mm between the first row of discs and the top of the sheet
- 2 mm between the first column of discs and the left side of the sheet.

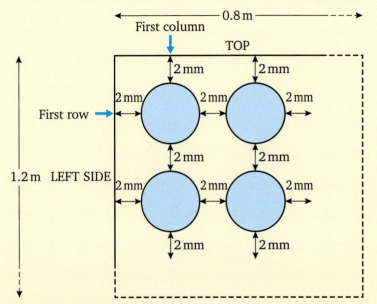

Not drawn accurately

a Show that 23 discs can be cut in each column.

b The unused part of a sheet is recycled.

What percentage of a sheet is recycled?

41 Use trial and improvement to find a solution to the equation $x^3 - 2x = 45$

The table shows the first trial.

x	$x^3 - 2x$	Conclusion
3	$3^3 - 2 \times 3 = 21$	Too small

Continue the table to find a solution to the equation.

Give your answer to one decimal place.

42 Adil and James are planning a mountain walk.

They find this rule to help them estimate how long the walk will take.

> **Estimating the time of a mountain walk**
> On a mountain walk it takes:
> 1 hour for every 3 miles travelled horizontally
> *plus*
> 1 hour for every 2000 feet climbed.
> Add 10 minutes of resting time for each hour you walk.

When planning their mountain walk, Adil and James estimate they will:

- travel 24 **kilometres** horizontally
- climb for 900 **metres**.

They plan to start their walk at 09:00.

At what time are they likely to complete their walk?

Hint

You will need to use these facts:
5 miles is approximately 8 km
3 metres is approximately 10 feet.

43 The diagram shows a scale drawing of Emma's bedroom.

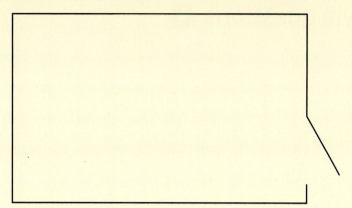

Scale : 1 cm represents 500 cm

Emma wants to paint the skirting boards in her bedroom.

Skirting boards are strips of wood at the bottom of the walls running around the edge of a room.

The skirting boards in her bedroom are 15 cm high.

Emma has a 1 litre tin of gloss paint.

According to the label, one litre of the paint covers between 10 and 12 square metres.

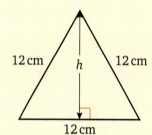

Does Emma have enough paint to paint the skirting boards with **two** coats of paint?

You must show working to justify your answer.

44 **a** The sides of an equilateral triangle are 12 centimetres long. Work out its height, h.

b A builder has a 2 metre long ladder.

He uses it to work on a wall and places the top of the ladder 1.8 metres from the ground.

To use the ladder safely the distance of the bottom of the ladder from the bottom of the wall, x, should be 0.45 metres.

Is the builder using the ladder safely?

Examination-style questions (k)

1   a   Complete the table of values for $y = x^2 - 4x$

x	-1	0	1	2	3	4	5
y		0	-3	-4		0	5

(2 marks)

   b   On a suitable grid, draw the graph of $y = x^2 - 4x$ for values of x from -1 to 5.   *(2 marks)*

   c   Use your graph to find the values of x *(2 marks)*

AQA 2008

2   Here are instructions for cooking a turkey.

> **Cook for 15 minutes at 220°C.**
>
> **Reduce the oven temperature to 160°C and cook for 40 minutes per kilogram.**

Kirsty is going to cook a 7 kilogram turkey.

She wants to take it out of the oven at 12. 45 pm.

At what time must she start to cook it?   *(4 marks)*

AQA 2009

Section 1

1 a $\frac{3}{4}$　**b** 9%　**c** 0.4　**d** 0.11

　e $\frac{3}{10}$ (other answers possible)

2 a never　　**c** sometimes　**e** sometimes
　b sometimes　**d** always　　**f** never

3 a

Tea	
Ordinary	
Apple	
Blackcurrant	
Lemon	
Other	

Key: = 4 cups

b

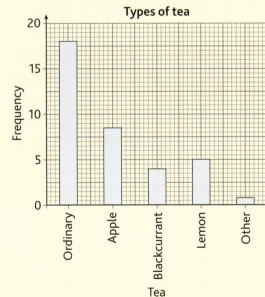

Types of tea

(bar chart, y-axis: Frequency 0–20, x-axis: Ordinary, Apple, Blackcurrant, Lemon, Other; Tea)

c

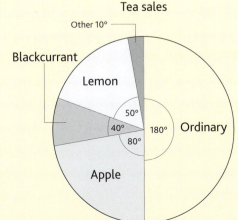

Tea sales

(pie chart: Other 10°, Blackcurrant, Lemon, 50°, 40°, 80°, 180° Ordinary, Apple)

4 a i $\frac{1}{6}$　**ii** $\frac{1}{5}$　　**b** £800

5 a $0.2 \times 0.4 = 0.08$　　**b** 20

6 £25 500

7 a 12

　b 2008 block drawn to 60

　c No, and explanation given (e.g. needs to build 84; another 17 needed but only 7 built in the previous year).

8 a No, and correct explanation (e.g. gives correct fraction $\frac{12}{30}$).

　b 2

9 49.28 seconds

10 5.9

11 The age of lions at London Zoo – quantitative and continuous

People's favourite cake at a show – qualitative

Points scored in 10 darts matches – quantitative and discrete

The weight of 12 newborn puppies – quantitative and continuous

The favourite building of people in Britain – qualitative

The average speed of train journeys into London – quantitative and continuous

The number of hours spent driving per day – quantitative and continuous

A person's shirt collar size – quantitative and discrete

The cost of bread – quantitative and discrete

Rainfall at a seaside resort – quantitative and continuous

12 30

13 a

Self-raising flour	480 g
Salt	1 teaspoon
Sultanas	150 g
Butter	80 g
Caster sugar	50 g
Egg	2 large
Milk	40 ml

b 600 g

c 48 : 15 : 8 : 5

14 a

	Wins for Andy	Wins for Roger
Grass Court	4	10
Hard Court	12	14

b Andy would prefer it to rain. Acceptable reason must refer to better ratio or percentage of wins on hard court or better probability.

15 a B

b 23

c comment referring to the trend (the more yellow cards, the more red) and to the exception of A

16 Any two of:

1G, 1B, 8R

2G, 2B, 6R

3G, 3B, 4R

17 a

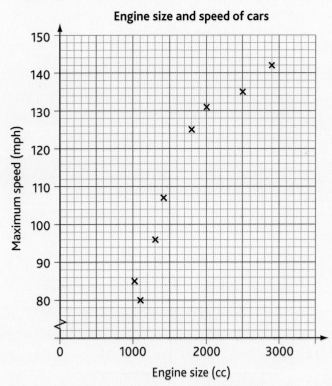

Engine size and speed of cars

b As a car's engine size increases, the maximum speed increases. This is positive correlation.

c

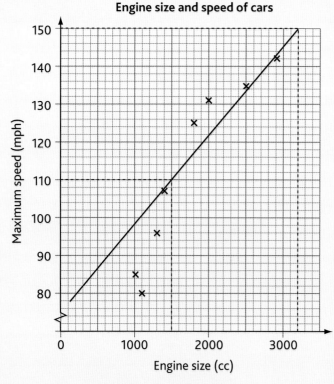

Engine size and speed of cars

i 110 (accept 98–102)

ii 3200 (accept 3100–3300)

d The point is beyond the limit of the data and thus the line of best fit is not very reliable.

18 70

19 77.6 cm.

20 a 13%   **b** 617

c No. More female members should be in the sample as there are more female members of the club. Sample should be selected on different days of the week. (Also, sample size is too small.)

21 5.12

22 a $\frac{44}{520} = \frac{11}{130}$

b picture card

23 Students' own diagram, which may be either a histogram or frequency polygon.

24 For Group A, sum 'fx' = 0 × 1 + 5 × 2 + 15 × 3 + 13 × 4 + 7 × 5 = 142

For Group B, sum 'fx' = 1 × 1 + 7 × 2 + 15 × 3 + a × 4 + b × 5 (where a and b are the missing frequencies)

= 60 + 4a + 5b

= 142 when a = 3 and b = 14

Students must also show that the sum of the frequencies is 40:

1 + 7 + 15 + a + b = 40 when a = 3 and b = 14

Examination-style questions

1 28.2

2 $\frac{1}{8}$

3 a i 14

    ii $\frac{5}{45} = \frac{1}{9}$

    iii £3

  b $\Sigma fx = (18 \times 6) + (10 \times 7) + (6 \times 8) + (4 \times 9) + (2 \times 10) = 282$

    $10 + 9 + 8 + 7 + 6 = 40$

    $\frac{282}{40} = 7.05$ so the mean hourly pay a Superspend is £7.05.

  c

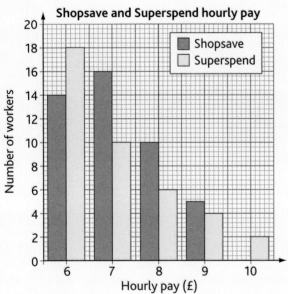

d i correct reason, e.g. some earn £10 at Superspend or higher relative frequency of £9

  ii correct reason, e.g. mean wage is higher at Shopsave or fewer earn £6

4 a Key: 49 | 1 represents 491 grams (or equivalent).

  b 501

  c Thirteen are overweight, which is above half, so the first claim is true. The second claim is not true because the limits are 490 and 510.

Section 2

1 a Dev   **b** Ben   **c** Alice

2 a

  b 12   **c** 6   **d** 42

3 a i 75 142   **ii** 12 547

  b i 900   **ii** 81

  c Three swaps are needed:

    1234 → 4231 (1↔4)

    4231 → 4132 (1↔2)

    4132 → 4312 (3↔1)

4 a and **f**

5 £0.15 or 15 pence

6 a Tim's number is 100 or 400 or 900

  b Stacey's number is 16

  c Afzal's number is 64

7 a Amy has 9 points, Bill has 17 points.

  b Amy must get at least 3 correct which scores 9 points giving a total of 18. So she needs at least 4 correct questions to get more than Bill.

8 $\frac{1}{6}$

9 a i 6.15 metres   **ii** £345.93

  b Four type A + four type C fit exactly.
  4 × 0.65 + 4 × 1.05 (= 6.8)

10 8 × 60 = 480
480 ÷ 100 = 4.8
5 tins

11 a B and C **c** C and D **e** A and B

  b B and D **d** B and E

12 a i $x = -14$
  ii $y = 7.5$
  iii $z = 0.9$

  b Sarah gets 90 votes

13 a $\frac{1}{2} + \frac{1}{3} = \frac{5}{6}$

  $\frac{1}{2} + \frac{1}{4} = \frac{3}{4}$

  $\frac{1}{3} + \frac{1}{4} = \frac{7}{12}$

  b i $\frac{1}{2} + \frac{1}{8}$
  ii Cut each loaf in half.
  Give each worker one of the halves.
  Cut each of the remaining two halves into quarters.
  Give each worker one of the quarters.

  c $\frac{1}{2} + \frac{1}{4}$
  Cut each loaf in half and give each worker one of the halves.
  Cut the remaining two halves in half and give the workers one piece each.

14 a $3x + 3y + 2x + 7y = 5x + 10y = 5(x + 2y)$

  b $2y - 2x$

  c $2y - 2x > 0$ so $2y > 2x$ so $y > x$
  Yes, she is right.

15 a 36 (79 − 43)

  b 2 × 4 + 3 = 11 (4th term of sequence A).
  4 × 3 − 1 = 11 (3rd term of sequence B).

16 45 red and 30 yellow

17 a $2^4 \times 3^2$

  b HCF = 6

  c LCM = 720

18

	A2B Cabs	Sapphire Taxis
Cinema	£58.60	£56.20
Exhibition	£51.20	£54.40
Show	£57	£63

So, the cheapest option will be to go to an exhibition and use A2B Cabs.

19 Option A: Cash payment of £388.80
Option B: £21.60 deposit, monthly payment of £68.40 = Total cost £432.00

20 $a = 4^9$, $b = 4^8$, $c = 4^7$
So, c is the smallest.

21 61

22 a $x > 7$ **b** −4, −3, −2, −1, 0, 1, 2

Examination-style questions

1 a 6

  b i £7.57
  ii Notes £5
  Coins £2, 50p, 5p, 2p

2 a 37

  b 9

3 a i 4.5 hours
  ii 0.75 hours
  iii 11.45 and 12.30 (steepest line)

  b 13.54

4 a $a = 2$

  b $b = 1$

  c $c = 2\frac{1}{2}$

Section 3

1 a shades 7 more triangles (3 left blank)
b 75%

2 a obtuse **b** right
c acute **d** reflex

3 No, she is £1.15 short.
Working: 2 ham & cheese sandwiches = £5.30
5 chicken sandwiches = £14.25
4 bags of crisps = £1.60
Total = £21.15

4 a pentagon **b** parallelogram

5 a -2 and $\frac{1}{2}$
b B (negative). Either a is $-$ve b is $+$ve c is $+$ve so ac is $-$ve or a is $+$ve b is $-$ve c is $-$ve so ac is $-$ve.

6 a 5 **c** (0, 2)
b (6, 0) **d** One point, (2, 4)

7 32 cm²

8 65°

9 12.8 cm

10 a £121.20 **b** 840 cm (or 8.4 m)

11 9.6 kg

12 80 cm

13 a $2d$ **b** -22 **c** $6g - 2f$

14 a 52% + 34% + 4% = 90% whereas the total should be 100%.
b i 0.419 km² **ii** 8%

15 36° or 45°

16 a 2855 **b** 400
c Germany. 1 euro is worth more pounds in Germany (150 euros = £128.76 in Germany, 150 euros = £127.99 in Britain)

17 a $17a + 7b$ **b** $a = 2$

18 a $x = 320, y = 55$ **b** $p = 132, r = 48$

19 a i 8 **ii** 20 **iii** 8000 **b** 20 000

20 A $y = 5$ B $y = x$ C $x = 3$ D $y = 2x$

21 $x = 55°, y = 35°$

22

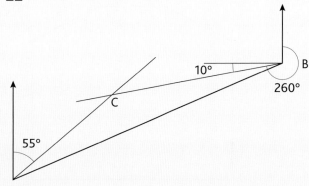

23 triangle (triangle = 336 cm², circle = 314.2 cm² to 1 d.p.)

24 ICT. 62.5% in ICT or ICT is 73.2 out of 120 in French

25 28

26 42 cm

27 a i Yes, they need 10 staff **ii** 1:7.6
b 6.1%

28 a $2x - 18$
b $3x - 2(4 - x)$

29 9.6

30 (10, 6)

31 120°

Examination-style questions

1 a 54° **b** 100° **c** 107°

2

Expression	Value
2x	8
5x	20
2x + 3y	5
y	−1
3x − y	13

Section 4

1 a $1\,cm^3$ **b** 19

2 a 1030

   b No, arrives 15 minutes late.

   c (no later than) 1545

3 a

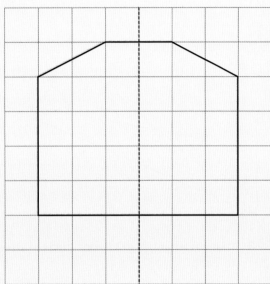

   b hexagon

4

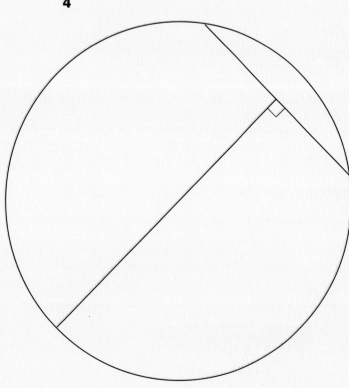

5

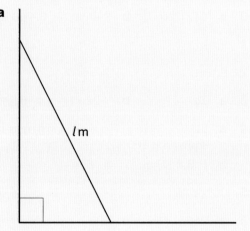

   b 11

6 a (2, 1)

   b i 7.0 cm to 7.2 cm   **ii** 30° to 32°

   c The lengths *AC* and *BC* are equal
       or
       Angles *CAB* and *CBA* are equal

7 a

lm

   b 4.9 m

8 a 3

   b

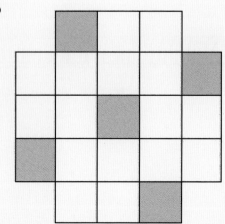

9 **a** Yes and a correct conversion factor stated or 320 km converted to 200 miles

   b 20 526

10 **a** £14

   b 5 days = £168, 6 days = £198 so not possible to get £178

11 $x = 40°, y = 65°, z = 50°$

12 10

13 34 cm

14 **a**

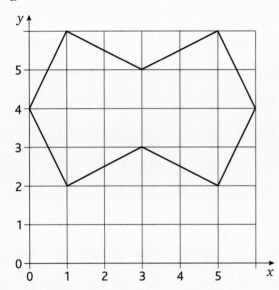

   b $y = 4$

15 All shapes have an area of 20 cm²

16 $\frac{1}{6}$

17 **a**

x	−1	0	1	2	3
y	−5	−3	−1	1	3

   b

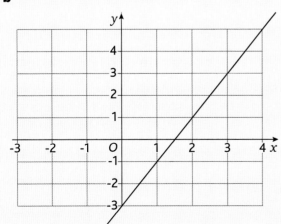

   c The solution of $2x − 3 = 0$ is where the graph intersects the x-axis.

18 **a**

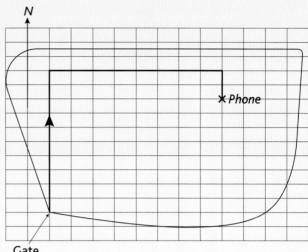

   b Move ___1___ metre ___East___

   Then move ___5___ metres ___South___

19 **a** 60 m   **b** 1.1 m   **c** 3.1 m

20 **a** **i**

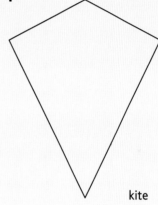

kite

   ii

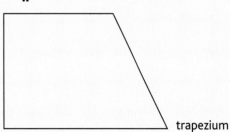

trapezium

   b Sum of angles in quadrilateral = 360°
   $360 − 3 × 90 = 90$
   So, if three angles are 90° the fourth angle must also be 90°

21 **a** trapezium

   b

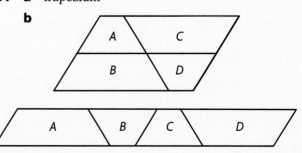

c 112.5°

d Using the outer quadrilateral:
67.5 + 67.5 = 135
67.5 + 135 + 67.5 = 270
$y = 360 - 270 = 90°$

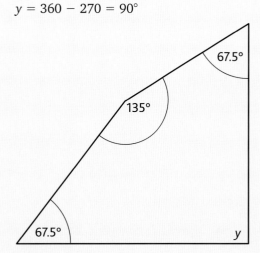

22 **a** $2x + 2$ or $2(x + 1)$

b $2x^2$

c $3x$

d $x^2 + 2$

23 **a** $BC = 3x + 2$

b $x + 4$

24 **a** Andrew 22 Pass
Bill 28 Pass
Clare 13 Fail

b 1 not attempted (-10)
3 correct $(+15)$
6 incorrect (-12)
$-10 + 15 - 12 = -7$

25

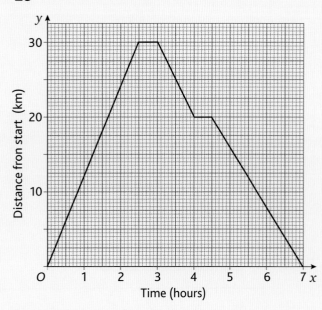

26 **a** and **b**

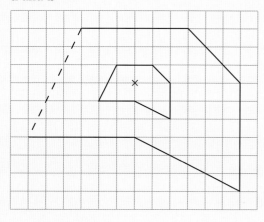

c 3

27 **a**

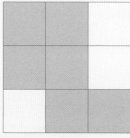

b

c

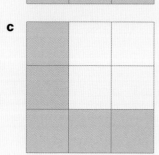

28 rotation 90° anticlockwise about O
translation 7 left 1 down

29 **a** 16 cm

b No. (Surface area is 376 cm² or explains
2 side faces of 5 × 6 (60 cm²) replace half
front and back faces of 6 × 8 (96 cm²) thus
reducing surface area by 36 cm²).

30 **a** $T = 12x + 8y$ **b** 44

31 **a** student's accurate drawing

b 87 m

32 a–c

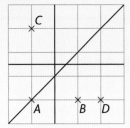

33 a

x	−3	−2	−1	0	1	2	3
y	14	9	6	5	6	9	14

b

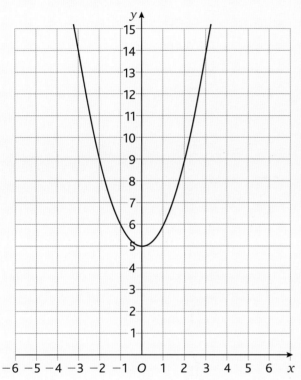

34

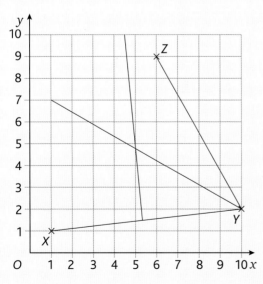

35 2.7 cm

36 No
24 ÷ 8 = 3
17.5 ÷ 5 = 3.5
Both of these scale factors need to be the same for the enlarged photo to fit exactly

37 a 96 cm³

b 6.6 cm with correctly filled in table with suitable values chosen

38 21 minutes (from 27–6)

39 6.5 cm

40 a 120 ÷ 5.2 = 23.08

b 29.4%

41 3.7 with correctly filled in table with suitable values chosen

42 Approximately 1630

43 Yes
Area that needs to be painted is
12.1 × 0.15 × 2 = 3.63 m²

44 a 10.39 cm (2 d.p.)

b If the ladder is 2 m long and the height is 1.8 m, then the ladder is 0.87 m from the wall so the builder is using the ladder safely.

Examination-style questions

1 a

x	−1	0	1	2	3	4	5
y	5	0	−3	−4	−3	0	5

b

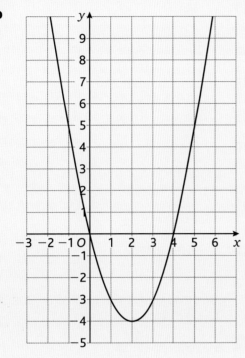

c The values of x for $y = 2$:
−0.5 to −0.4 and 4.4 to 4.5.

2 7.50am